How Gary Glucose Became ATP

A Journey into the Kreb's Cycle

Written and Illustrated

by Sara Brandow

Copyright © 2017 Sara Brandow

All rights reserved.

ISBN-10: 1981471111

ISBN-13: 978-1981471119

DEDICATION

This book is dedicated to every student (like me) that ever struggled to understand biology, in particular the Kreb's Cycle. For years, I struggled to understand the Kreb's Cycle and wished someone would present this complicated process in storybook form because it seemed it would make the whole concept easier to comprehend and mentally visualize.

I went ahead and made that wish a reality, and I hope you enjoy this story while at the same time it helps you understand some important details of the Kreb's Cycle.

Please enjoy this book and think of Gary Glucose whenever you recall how ATP is formed!

CONTENTS

	Acknowledgments	i
1	Gary Glucose Becomes 36-38 ATP	1
2	Glenn Glucose Becomes 2 ATP	14
3	Gabriella Glucose Review Game	18

ACKNOWLEDGMENTS

Thank you, Professor Hernandez for finally making biology make sense to me and igniting my passion for a subject I thought I would never understand.

My husband deserves tremendous thanks for believing in me, encouraging me to pursue my dreams, and sticking by me when the going got tough. Thank you, Brian!

And to my parents who taught my that my stubbornness was really just tenacity and I could overcome any obstacle, praise to you!

GARY GLUCOSE BECOMES 36-38 ATP

Once upon a time, there was a glucose molecule named Gary who wanted nothing more than to become energy in the form of 36-38 ATP.

He was very sad that he was not ATP, until one day he met Phosphofructokinase (PFK for short) on a certain path named "Glycolysis." PFK helped him become a new substance called Glyceraldehyde-3-phospate, also known as "PGAL." As soon as Gary had become this PGAL substance, PFK disappeared, nowhere to be seen.

HOW GARY GLUCOSE BECAME ATP

Gary continued walking down the Glycolysis pathway.

While he was on the Glycolysis Trail, he had once again transformed into a new substance called Pyruvate by a molecule of NADH. NADH was friendly, but he seemed to be in a great hurry with no time to talk. NADH went off into the woods and promised he would be back again soon to help Gary Glucose become ATP.

Soon, Gary came to a house along the path and he knew he needed to go inside. The sign on the house said "Krebs Cycle" and there was a sign in the front yard saying that Oxygen was home, so he could go inside.

Gary walked up the sidewalk to the house. While we walked on the sidewalk, he stepped in a puddle of Coenzyme A and got it stuck on his shoe! He also noticed that he lost a carbon dioxide molecule. Again, NADH seemed to come out of nowhere and run into Gary and then leave again, promising he would be back to help Gary become ATP. All of these things had turned him into Acetyl CoA!

When Gary made it to the front door, the door opened right up for him so he could go inside.

When Gary went inside the Kreb's Cycle house, he met some new friends – one of them was someone called Oxaloacetate.

Oxaloacetate took Gary by the arm and helped him transform into Citrate. Soon, Gary was engulfed in a cloud and surrounded by lots of other molecules whom he had never seen before.

He was mesmerized by these incredible new molecules! When he walked past a mirror, he didn't even recognize himself because he had changed so much!

Then some molecules that looked a lot like the friend he made earlier, NADH, bumped into him and ran to somewhere else in the house. As always, they promised to be back again soon and help him become ATP. Another molecule that looked similar to NADH also bumped into him. This molecule said his name was FADH. Like the NADH, he said he would be back to help him a little later.

Gary was beginning to get a little scared. He wasn't sure he wanted to become ATP after all and all of these new molecules were crowding around him and changing him into this and that. He had no control over what was happening, so he decided to wait and see what happened. He was sure he would be alright.

Besides, Olivia Oxygen seemed like a nice molecule and he trusted her.

Olivia Oxygen then took Gary Glucose down the stairs into the basement. "Oh, no!" Gary thought. "Maybe I shouldn't have been so trusting after all!" What was she doing? To his surprise, his friends NADH and FADH were also there. He watched them as they played on a series of four bounce-houses that were labeled "Electron Transport Chain." It was mesmerizing to watch!

As they bounced, Hydrogen atoms were released into the air and then the Hydrogen atoms fell back down to the floor through an interesting contraption labeled "ATP Synthase." Out of the contraption came ATP! Gary had become approximately 38 ATP through the interactions with his friends along the way and after this final step! Hooray!

...BUT WAIT!

THE STORY ISN'T OVER YET!...

GLENN GLUCOSE BECOMES 2 ATP

Gary was so happy to become ATP that his best friend Glenn wanted to become ATP too! Glenn went down the same path called "Glycolysis" and he saw the "Kreb's Cycle" house. He knocked on the door to come inside – but Olivia Oxygen was not home. There was a note on the door saying that if ATP was needed immediately, he could take a yeast plant from her garden and use it to ferment. So Glenn did what the note said and found a yeast plant.

HOW GARY GLUCOSE BECAME ATP

As he fermented, he became ethanol, CO_2, ATP, and heat while he sat on her doorstep. Granted, he only got to become 2 ATP, but he did get a bonus! He got to blow more CO_2 bubbles than Gary did!

Now that they were both energetic ATP molecules, Gary and Glenn decided what to do with their newfound energy.

Gary thought he would help his friend Sally the Snail respire. Glenn thought it would be fun to help bread rise or maybe ferment some wine. In the end, they were both very happy with what they had accomplished.

THE END

Gabriella Glucose Review Game

To review what you have learned, please follow the instructions on the next page! You may need to refer to your textbook, but get your colored pencils out and have fun! Feel free to make copies and share. Check your answers on the answer key at the end of this book.

HOW GARY GLUCOSE BECAME ATP

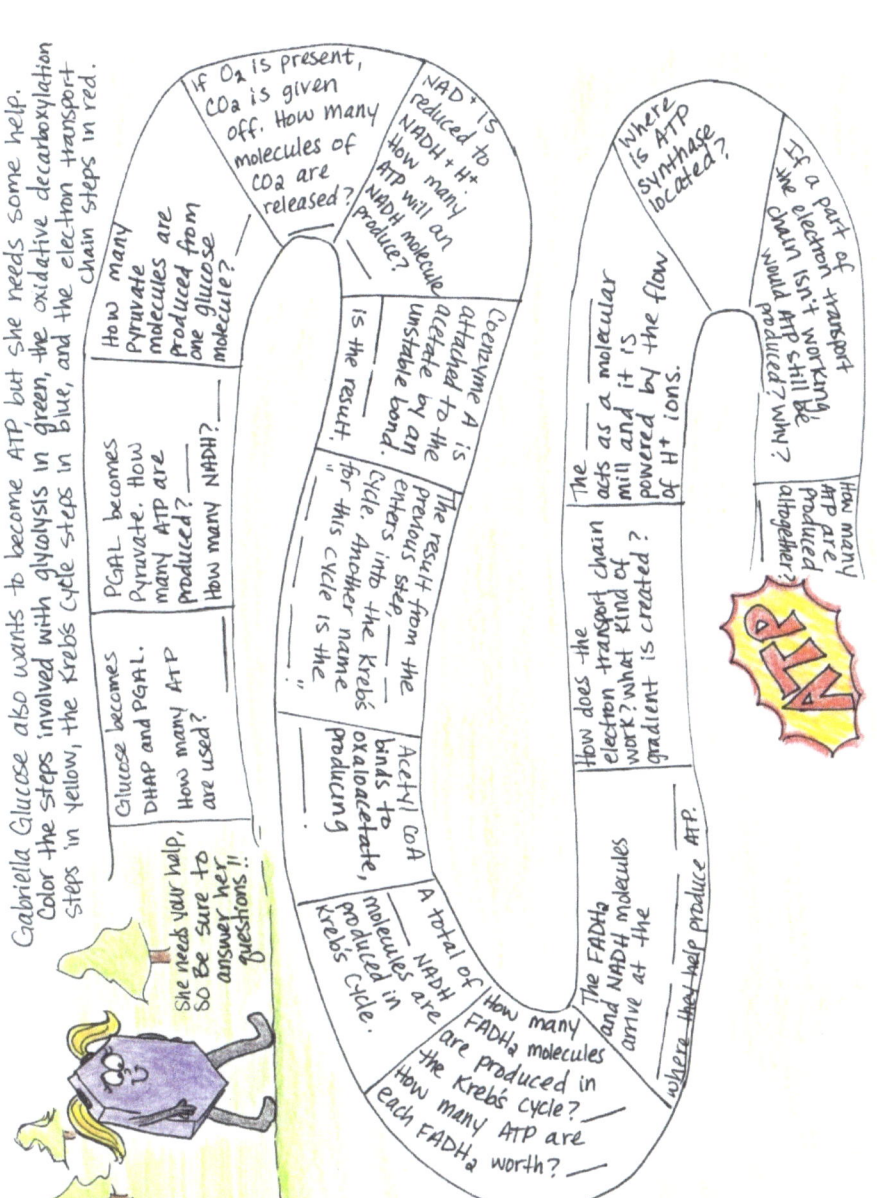

Answer Key:

1. Glucose becomes Dihydroxyacetone phosphate (DHAP) and Glyceraldehyde-3-phosphate (PGAL) during glycolysis. 2 ATP are used in this phase.
2. PGAL becomes pyruvate in the glycolysis phase, and 4 ATP are produced. 2 NADH are produced.
3. Two pyruvate molecules are produced from one glucose molecule during glycolysis.
4. When O2 is present, a TOTAL of 3 CO2 molecules will be given off during the entire process: one CO2 molecule during the oxidative decarboxylation phase and two CO2 molecules during the Kreb's/Citric Acid Cycle.
5. A molecule of NADH will produce a maximum of approximately 3 ATP during the processes of the electron transport chain.
6. Coenzyme A is attached to the acetate by an unstable bond; acetyle CoA is the result (this occurs between the stages of glycolysis and the Kreb's Cycle).
7. Another name for the Kreb's Cycle is the Citric Acid Cycle.
8. When Acetyl CoA binds to oxaloacetate within the Kreb's Cycle, citrate is formed.
9. Three NADH molecules are formed during Kreb's Cycle
10. Only one $FADH_2$ molecule is producd in Kreb's Cycle but it is worth 2 ATP.
11. $FADH_2$ and NADH arrive at the electron transport chain where they help produce ATP.
12. The electron transport chain works by pumping H+ protons. This creates a H+ gradient across the inner mitochondrial membrane.
13. Powered by the flow of H+ ions, ATP synthase acts as a molecular mill.
14. ATP synthase is located in mitochondrial and chloroplast membranes or eukaryotes and in the plasma membrane of prokaryotes.
15. If a part of the electron transport chain isn't working, ATP would still be produced, but at a much slower rate.
16. 36-38 ATP are produced per molecule of glucose.

ABOUT THE AUTHOR

Sara Brandow works as a licensed physical therapist assistant in Colorado Springs, CO. She has always had a passion for learning and believes change and challenge are essential parts of growth. Biology was once a challenging subject for her, but with persistence, prayer, determination, and the right instructor, it finally made sense. If you struggle with biology (or any subject), please don't give up!

Sara loves animals, creating art, experiencing nature, and helping others. It is her hope that this book will clarify the Kreb's Cycle for her readers and perhaps make the whole topic a bit more fun!

www.ingramcontent.com/pod-product-compliance
Lightning Source LLC
Chambersburg PA
CBHW040056250526
45473CB00042B/2783